Universal Voting System

Empowering Decision Making

Dan Paul

Contents

Chapter 1

Proposal for a Universal Voting System

Executive Summary

This proposal outlines the design, implementation, and funding of a Universal Voting System (UVS). The UVS will allow individuals to securely vote on any issue, anytime, anywhere. The system will utilize biometric authentication, including thumbprint and facial recognition, and provide an accessible platform for individuals, governments, and organizations to upload ballots for public voting. This proposal highlights the system's potential for global governance, community engagement, and technological innovation, alongside a detailed timeline, budget, and incentive structure for widespread adoption.

Vision

The UVS aims to create a platform that empowers citizens worldwide to participate in decision-making processes. From neighborhood issues to international policies, the system democratizes access to voting and fosters a culture of transparency and collaboration.

System Specifications

Core Features:

1 Biometric Authentication:

○ Thumbprint and facial recognition to ensure secure access.

○ Encrypted data storage to protect personal information.

2 Multi-Level Ballot Access:

○ Ballots for local, state, national, and international issues.

○ Ability for individuals, organizations, and governments to upload ballots.

3 User-Friendly Interface:

○ Intuitive design for all age groups and literacy levels.

○ Language support for all major languages.

4 Real-Time Voting:

○ Immediate results displayed after the voting period ends.

○ Transparent vote-counting processes.

5 Accessibility Features:

○ Compatibility with screen readers and other assistive technologies.

○ Offline options for rural areas using portable voting kiosks.

6 Audit and Oversight:

○ Blockchain-based vote tracking for transparency and security.

○ Independent monitoring bodies to ensure integrity.

Implementation Plan

Phase 1: Research and Development (Year 1)

- **Tasks:**

○ Develop and test biometric authentication technology.

○ Create the initial software prototype for the UVS platform.

○ Consult with cybersecurity experts to ensure data protection.

○ Engage with stakeholders to gather requirements and feedback.

- **Budget:** $50 million.

Phase 2: Pilot Testing (Year 2)

- **Tasks:**

○ Deploy pilot systems in five diverse regions (urban, rural, developed, developing).

○ Train local facilitators and gather user feedback.

○ Refine the system based on pilot results.

- **Budget:** $100 million.

Phase 3: Nationwide Rollout (Years 3-4)

- **Tasks:**

○ Collaborate with national governments for implementation.

○ Ensure all computers and kiosks are equipped with UVS technology.

○ Launch public awareness campaigns to promote adoption.

- **Budget:** $1 billion.

Phase 4: Global Expansion (Years 5-7)

- **Tasks:**

○ Partner with international organizations to introduce UVS globally.

○ Establish regional data centers for efficient operation.

○ Expand language and cultural customization features.

- **Budget:** $5 billion.

Total Estimated Budget: $6.15 billion over 7 years.

Incentives for Adoption

For Individuals:

- Earn points redeemable for goods, services, or tax benefits by participating in votes.
- Gamified system to encourage civic engagement, with badges and recognition for regular voters.

For Governments and Organizations:

- Access to anonymized analytics for better policymaking.
- Streamlined public opinion gathering for urgent decisions.
- Funding grants for local governments adopting UVS early.

For Ballot Creators:

- Crowdfunding options directly linked to ballot initiatives.
- Recognition and sponsorship opportunities for successful initiatives.

. . .

Use Cases

Governments:

• National elections, referendums, and policy decisions.

• Local governance issues like zoning and community projects.

Groups and Organizations:

• Nonprofits: Decision-making on resource allocation.

• Corporations: Employee voting on workplace policies.

• Community Groups: Votes on local events and initiatives.

Individuals:

• Creating grassroots movements by uploading petitions and polls.

• Engaging friends and family in collaborative decision-making.

Potential Impact

1 Increased Voter Participation:

○ Accessibility and convenience encourage higher turnout.

○ Continuous voting opportunities foster habitual civic engagement.

2 Transparency and Trust:

○ Blockchain and biometric verification ensure secure and transparent processes.

3 Global Collaboration:

○ A platform for addressing global issues such as climate change and human rights.

Call to Action

To realize this transformative vision, we invite billionaires, investors, and governments to join this initiative. By funding the UVS, you will contribute to a more democratic, inclusive, and transparent world. Let's shape the future of decision-making together.

Chapter 2

Universal Voting Systems – Lessons from Government Initiatives

Introduction

As governments worldwide experiment with digital and universal voting systems, their experiences offer valuable insights into the challenges and opportunities such systems present. This chapter examines the efforts of Estonia, Switzerland, the United States (specifically West Virginia), and China, analyzing how they funded their systems, the obstacles they faced, and the innovative solutions they employed. Finally, it explores how a combined approach, leveraging the best elements of each, could create a truly universal voting system.

Estonia: The Pioneer of i-Voting

Funding and Development Timeline:

• **Timeline:** The Estonian i-Voting system launched in 2005 as part of a broader e-governance initiative.

• **Funding:** Public funds, supplemented by European Union (EU) grants, financed the system's development. The total cost of initial implementation was approximately $10 million.

Achievements:

• Estonia became the first country to allow nationwide internet voting in general elections.

• By 2019, over 40% of voters used i-Voting in national elections.

Challenges:

1 Security Concerns: Early fears of hacking and vote manipulation prompted the government to invest heavily in cybersecurity measures, including end-to-end encryption and independent audits.

2 Public Trust: Skepticism among older voters was addressed through education campaigns and transparent processes.

3 Scalability: Estonia's small population (1.3 million) made the system feasible, but scalability to larger nations remains untested.

Support and Opposition:

• **Support:** Younger generations, tech-savvy citizens, and progressive political parties championed the system.

• **Opposition:** Critics raised concerns about digital exclusion for those without internet access.

Switzerland: Blockchain Voting Pilots

Funding and Development Timeline:

• **Timeline:** Blockchain-based voting pilots began in 2016 and ran for several years.

• **Funding:** Joint efforts between local governments and blockchain technology startups funded the pilots, with costs ranging from $1-5 million per pilot.

Achievements:

• Demonstrated the potential of blockchain technology for secure, transparent, and auditable elections.

• Pilots in Geneva and Zug gained positive feedback for their user-friendly interfaces.

Challenges:

1 **Technical Complexity:** Blockchain systems required specialized infrastructure and expertise, leading to high costs.

2 **Adoption Reluctance:** Many voters were unfamiliar with blockchain, reducing participation rates during pilot programs.

Support and Opposition:

• **Support:** Proponents highlighted blockchain's transparency and resistance to tampering.

• **Opposition:** Critics argued that the technology was overengineered and not cost-effective for small-scale elections.

United States: West Virginia Blockchain Voting

Funding and Development Timeline:

• **Timeline:** Blockchain voting was introduced in West Virginia for absentee ballots during the 2018 midterm elections and continued through 2020.

• **Funding:** The initiative was funded by Tusk Philanthropies and state resources, with costs exceeding $1 million.

Achievements:

• Enabled military personnel stationed overseas to vote securely.

• Increased accessibility for absentee voters in rural areas.

Challenges:

1 Security and Privacy: Concerns about blockchain's ability to anonymize votes led to criticism from cybersecurity experts.

2 Limited Scope: The system was only available to a small subset of voters, limiting its impact.

Support and Opposition:

• **Support:** Military voters and their families appreciated the convenience.

• **Opposition:** Election security experts questioned the system's scalability and resistance to sophisticated cyberattacks.

China: Biometric Voting Systems

Funding and Development Timeline:

• **Timeline:** Biometric voting systems were rolled out as part of a broader digital governance initiative starting in the early 2010s.

• **Funding:** Fully funded by the central government as part of its smart city projects.

Achievements:

• Widespread adoption of thumbprint and facial recognition technologies.

• Efficient local governance and streamlined voting processes for internal community decisions.

Challenges:

1 **Transparency:** The centralized nature of China's system raised concerns about vote manipulation and lack of independent oversight.

2 Surveillance Fears: Critics argued that biometric data collection could be misused for surveillance purposes rather than democratic engagement.

Support and Opposition:

- **Support:** Local governments and state-run organizations benefited from the system's efficiency.

- **Opposition:** International human rights groups raised alarms about privacy violations.

Toward a Universal Voting System

Combining the Best Features:

1 Security from Estonia: Implement end-to-end encryption and independent audits to build trust.

2 Transparency from Blockchain Pilots: Use blockchain to ensure vote immutability and enable real-time audits.

3 Accessibility from West Virginia: Develop mobile-friendly platforms for absentee and rural voters.

4 Efficiency from China: Leverage biometric authentication for seamless voter verification.

Key Considerations:

1 Funding:

○ Blend public funding with private partnerships and philanthropic contributions.

○ Estimated cost for global implementation: $6-10 billion over 5-7 years.

2 Technical Challenges:

○ Scalability: Use cloud-based infrastructure to accommodate large populations.

○ Cybersecurity: Establish a global task force to monitor and mitigate threats.

3 Public Trust:

○ Transparency: Publish open-source code and engage independent monitors.

○ Education: Launch campaigns to familiarize voters with the system.

Potential Timeline:

- **Year 1:** Research and development.
- **Years 2-3:** Pilot programs in diverse regions.
- **Years 4-5:** Nationwide rollouts in participating countries.
- **Years 6-7:** Global expansion and system refinement.

Conclusion

The experiences of Estonia, Switzerland, the United States, and China highlight both the promise and pitfalls of digital voting systems. By learning from these initiatives and combining their strengths, a Universal Voting System can become a reality, fostering global democratic engagement and creating a future where every voice truly matters.

Chapter 3

Universal Voting Systems – Benefits, Drawbacks, and Best Practices

Introduction

A universal voting system promises to revolutionize democratic participation by enabling seamless, accessible, and inclusive voting processes for various groups. However, the implementation of such a system requires careful consideration of the unique benefits and drawbacks it presents for small groups, community organizations, and governments at local, state, national, and international levels. This chapter explores these aspects and provides a set of dos and don'ts for creating an effective and trustworthy universal voting system.

Benefits and Drawbacks Across Governance Levels

Small Groups and Community Organizations

Benefits:

1 Increased Participation: A universal voting system enables members to vote conveniently, increasing engagement.

2 Streamlined Decision-Making: Real-time vote collection and analysis allow for quicker resolutions on community issues.

3 Transparency: Digital records of votes improve accountability within the group.

Drawbacks:

1 Technical Barriers: Some members may lack the digital literacy required to use the system effectively.

2 Cost: For smaller groups, implementing and maintaining the system might be financially challenging.

Local Governments

Benefits:

1 Enhanced Accessibility: Residents can participate in local elections from anywhere, boosting voter turnout.

2 Cost Efficiency: Reduces the expenses associated with physical polling stations and paper ballots.

3 Real-Time Feedback: Enables local governments to quickly gauge public opinion on urgent issues.

Drawbacks:

1 Security Concerns: Limited budgets may hinder the ability to implement robust cybersecurity measures.

2 Digital Divide: Lower-income or rural areas may face challenges due to limited internet access.

City Governments

Benefits:

1 Civic Engagement: Encourages greater participation in city-level governance, such as budgeting or zoning decisions.

2 Efficient Resource Allocation: Voting data can help city officials allocate resources more effectively.

3 Scalable Systems: Cities can adapt universal voting systems to accommodate diverse populations.

Drawbacks:

1 Infrastructure Challenges: Larger cities may face difficulties in scaling the system for millions of residents.

2 Opposition to Change: Traditionalists may resist transitioning from paper ballots to digital voting.

State Governments

Benefits:

1 Uniformity: Standardized systems across municipalities ensure consistency and reduce administrative errors.

2 Inclusivity: Allows absentee voters, such as military personnel, to participate more easily.

3 Cost Savings: Eliminates the need for extensive physical infrastructure during elections.

Drawbacks:

1 Cybersecurity Risks: Larger systems become more attractive targets for hackers.

2 Complexity: Coordinating between state and local governments may lead to logistical challenges.

National Governments

Benefits:

1 Increased Turnout: National elections become more accessible, especially for overseas citizens.

2 Data-Driven Insights: Enables governments to analyze voter trends and make informed policy decisions.

3 Environmental Impact: Reduces the carbon footprint associated with traditional voting methods.

Drawbacks:

1 Polarization Risks: Disinformation campaigns could undermine trust in the system.

2 High Costs: Initial implementation and cybersecurity expenses can be significant.

International Governance and Organizations

Benefits:

1 Global Participation: Allows stakeholders from multiple nations to have a voice in global decision-making processes.

2 Standardization: Creates a unified approach to addressing global challenges, such as climate change.

3 Transparency: Ensures that international decisions are made with broad and diverse input.

Drawbacks:

1 Cultural and Legal Differences: Variations in laws and customs may complicate implementation.

2 Language Barriers: Multilingual interfaces require careful design to ensure inclusivity.

3 Trust Issues: Skepticism about fairness and neutrality in international voting could hinder adoption.

• • •

Dos and Don'ts for Universal Voting Systems

Dos:

1 Prioritize Security: Implement end-to-end encryption, biometric verification, and third-party audits to safeguard the system.

2 Ensure Accessibility: Design user-friendly interfaces and provide offline options for regions with limited internet access.

3 Promote Transparency: Make source codes open for independent review and publish results in real-time.

4 Educate the Public: Launch awareness campaigns to familiarize voters with the system's features and benefits.

5 Test Extensively: Conduct pilots and stress tests to identify and address vulnerabilities before full implementation.

Don'ts:

1 Ignore Inclusivity: Avoid designing systems that exclude individuals without advanced digital skills or access to technology.

2 Compromise on Privacy: Ensure that biometric and personal data are securely stored and used solely for verification purposes.

3 Underestimate Costs: Factor in ongoing expenses for maintenance, upgrades, and cybersecurity.

4 Rush Implementation: Avoid skipping critical testing phases to meet arbitrary deadlines.

5 Disregard Feedback: Continuously seek input from stakeholders, including voters, experts, and advocacy groups.

Conclusion

A universal voting system holds immense potential to enhance democratic participation across all levels of governance. By understanding the unique needs and challenges of small groups, communities, and governments, and adhering to best practices, such a system can foster trust, efficiency, and inclusivity. However, its success depends on careful planning, robust security, and unwavering commitment to transparency and accessibility.

Chapter 4

The Opposition to a Universal Voting System

A universal voting system, where anyone can vote on anything, anywhere, at any time, represents an ambitious vision of democratizing decision-making on an unprecedented scale. Such a system promises accessibility, inclusivity, and transparency. However, its implementation would face resistance, not only due to technical and logistical hurdles but also because of entrenched political interests on both sides of the aisle. To understand the opposition, we must explore how current political parties in the United States, specifically Republicans and Democrats, approach voting restrictions, and what a universal voting system might mean for disadvantaged populations.

How Republicans Restrict Voting

Republicans often frame their voting restrictions as measures to combat voter fraud and ensure election integrity. However, critics argue that these policies disproportionately impact marginalized groups and suppress voter turnout among demographics that tend to vote Democratic, such as minorities, low-income individuals, and young people. Common Republican-supported restrictions include:

• **Voter ID Laws**: Requiring government-issued photo identification to vote. While proponents argue this prevents fraud, critics point out that many individuals, especially low-income and elderly citizens, lack access to such IDs.

• **Polling Place Closures**: Particularly in rural or minority-heavy areas, reducing the number of polling places leads to long lines and discourages turnout.

• **Reduced Early Voting**: Limiting early voting windows makes it harder for people with rigid work schedules to cast their ballots.

• **Restrictions on Mail-In Voting**: Tightening the rules for absentee and mail-in ballots can disproportionately affect voters with disabilities, the elderly, and those living in remote areas.

• **Voter Roll Purges**: Aggressive removal of voters from registration lists often mistakenly disenfranchises eligible voters.

How Democrats Restrict Voting

While Democrats generally advocate for expanded voting access, they are not without criticism regarding their approach to elections. Critics accuse them of engaging in restrictive practices that serve to consolidate power within their own political strongholds. Examples include:

- **Manipulation of Primary Processes**: Superdelegates and closed primaries can limit the influence of grassroots candidates and independent voters, favoring establishment figures.

- **Ballot Access Restrictions for Third Parties**: Democrats, along with Republicans, have historically supported high signature thresholds and filing fees that make it difficult for third-party candidates to qualify for ballots.

- **Gerrymandering**: While often associated with Republicans, Democrats in states they control have also used gerrymandering to maintain political dominance, diluting the voting power of opposition groups.

Impact of a Universal Voting System on Voter Turnout

A universal voting system could radically reshape the voting landscape by removing many of the barriers that currently discourage or prevent participation. By allowing people to vote anytime, anywhere, and on any issue, such a system would likely increase turnout, particularly among those who face significant obstacles under the current framework.

- **Accessibility**: Digital voting platforms could enable voters to cast their ballots from home or mobile devices, making participation far easier for people with disabilities, transportation challenges, or inflexible work schedules.

• **Elimination of Geographic Barriers**: Voters would no longer be restricted to polling places, removing the impact of polling closures or unequal distribution of resources.

• **Real-Time Voting Opportunities**: If voting could occur in real time on a wide range of issues, individuals might feel more engaged and empowered to participate.

Impact on Disadvantaged Populations

Universal voting could be transformative for individuals who have historically faced barriers to voting:

• **People with Disabilities**: Accessible interfaces and remote voting options would eliminate many physical obstacles.

• **Low-Income Voters**: Eliminating transportation costs and offering more flexible voting times would reduce barriers related to financial insecurity.

• **Non-Native English Speakers**: Multilingual support within a universal system could better serve immigrant populations.

• **Formerly Incarcerated Individuals**: In a truly universal system, voting rights could be decoupled from criminal records, enfranchising millions who are currently disenfranchised.

- **Young Voters**: Digital platforms align with the habits of younger generations, encouraging their participation.

Challenges and Critiques of a Universal Voting System

Despite its potential benefits, a universal voting system would face significant opposition. Critics might argue that such a system is vulnerable to fraud, cybersecurity threats, and coercion. Political parties might resist losing control over the mechanisms of elections, fearing a loss of influence. Additionally, questions of equity might arise if access to digital tools remains uneven.

However, the most profound opposition would likely come from those who benefit from the status quo. A universal voting system challenges power structures by expanding access to historically disenfranchised groups. The fear for some is that broader participation would lead to outcomes that upend the established order.

Conclusion

The opposition to a universal voting system is deeply rooted in the struggle for power and influence. Both Republicans and Democrats have, at times, supported measures that restrict access to voting to maintain their political advantages. A universal system, however, offers the potential to level the playing field, empowering marginalized voices and creating a more participatory democracy. The key challenge lies in balancing security, accessibility, and equity while addressing the entrenched interests that benefit from limiting participation. If successful, a universal voting system could fundamentally transform how we think about democracy itself.

Chapter 5

The Impact of a Universal Voting System

A universal voting system, where anyone can vote anywhere, anytime, and on any issue, represents a paradigm shift in the democratic process. By radically democratizing decision-making, this system has the potential to reshape not only governance but also the economic, social, and environmental fabric of society. This chapter examines the local, state, and national impacts of such a system, exploring its consequences for policy, ecosystems, and the people who inhabit them.

Local Impact: Grassroots Empowerment and Community Decision-Making

At the local level, a universal voting system could foster unprecedented civic engagement. Communities would have the ability to vote on matters ranging from zoning laws to school budgets in real time. This hyper-local empowerment could lead to:

• **Enhanced Civic Participation**: With easier access to voting, more individuals could engage in shaping their immediate surroundings, leading to policies that better reflect local needs.

• **Economic Revitalization**: Communities might vote to prioritize small businesses, green spaces, and public transportation projects, stimulating local economies and improving quality of life.

• **Environmental Stewardship**: Local residents could directly influence decisions about land use, water management, and waste reduction, leading to more sustainable practices.

However, challenges may arise from the sheer volume of decisions requiring voter input. Communities might need to adopt educational initiatives to ensure informed participation, and safeguards would be necessary to prevent majority rule from marginalizing minority voices.

Statewide Impact: Bridging Divides and Broadening Representation

State governments often serve as intermediaries between local and national interests. A universal voting system would transform this role, enabling more direct input from residents on state-level decisions. Potential impacts include:

• **Equitable Resource Distribution**: Residents could vote on how state budgets are allocated, ensuring that rural and urban areas receive fair representation in funding.

• **Addressing Regional Inequalities**: States could better tailor policies to reflect the diverse needs of their populations, from education reform to healthcare access.

• **Decentralized Decision-Making**: Statewide referendums could become more frequent, allowing citizens to vote on pressing issues without waiting for legislative action.

This system could also expose and exacerbate regional divides, as areas with differing priorities might clash over state-level decisions. Collaborative platforms would be needed to mediate these conflicts and promote compromise.

National Impact: Redefining Democracy and Policy

At the national level, a universal voting system could radically transform governance. Citizens would have a direct say in issues ranging from foreign policy to federal budgets. The implications for the nation include:

• **Economic Transformation**: Voters might prioritize funding for renewable energy, universal healthcare, and infrastructure improvements, potentially reshaping the economy toward more sustainable and equitable models.

• **Social Justice Advancements**: Policies addressing systemic inequality could gain broader support, as marginalized communities would have a stronger voice in national debates.

• **Political Realignment**: Traditional party structures might weaken, as voters bypass intermediaries to directly influence policy. This could lead to a more issue-driven political landscape.

The challenges at this scale are immense. National decisions often require deep expertise and long-term planning, which may not align with the preferences of a majority voting in the short term. Balancing populist impulses with the need for strategic governance would be a critical hurdle.

Environmental Consequences: Air, Land, Water, and Ecosystems

A universal voting system could have profound implications for the natural world. With greater public input, policies might shift to prioritize environmental sustainability:

• **Air Quality**: Voters could support stricter regulations on emissions and incentives for clean energy, reducing air pollution and combating climate change.

• **Land Use**: Decisions on urban development, agriculture, and conservation could reflect broader societal values, protecting natural habitats and promoting biodiversity.

• **Water Management**: Public input could lead to investments in clean water infrastructure and stricter regulations on pollution, benefiting both human health and ecosystems.

On the other hand, the system could also produce unintended consequences. For instance, if voters prioritize short-term economic gains over environmental preservation, ecosystems could suffer.

Social and Cultural Shifts: People and Communities

The social implications of a universal voting system are vast. By giving everyone a voice, the system could:

• **Empower Marginalized Groups**: Individuals who have historically been excluded from the democratic process, such as low-income citizens, immigrants, and the disabled, would gain greater influence.

• **Foster Community Cohesion**: Shared decision-making could strengthen social bonds and promote a sense of collective responsibility.

• **Challenge Social Norms**: As diverse voices gain prominence, societal attitudes toward issues like gender equality, racial justice, and LGBTQ+ rights could evolve more rapidly.

However, the system could also exacerbate divisions. Polarized opinions might lead to gridlock or social fragmentation, especially on contentious issues.

Winners and Losers: Who Gains and Who Loses?

A universal voting system would create both opportunities and challenges, redistributing power in ways that benefit some while disadvantaging others:

- **Winners**: Disenfranchised communities, grassroots organizations, and proponents of progressive policies would likely gain influence. Environmentalists, educators, and public health advocates might also see their priorities advanced.

- **Losers**: Established political elites, corporate interests, and groups reliant on maintaining the status quo could lose power. Industries resistant to change, such as fossil fuels, might face significant challenges.

The ultimate impact of a universal voting system depends on how it is designed and implemented. While it offers the potential for a more inclusive and equitable society, careful planning is essential to avoid unintended consequences and ensure that the system truly serves the public good.

Conclusion

A universal voting system represents a bold step toward a more participatory democracy. By enabling anyone to vote on anything, anywhere, at any time, it has the potential to transform decision-making at every level. From local neighborhoods to the national stage, the impacts on economic, social, political, and environmental policies could be profound. However, realizing this vision requires navigating significant challenges, balancing competing interests, and ensuring that the system works for all—humans, animals, and ecosystems alike. The stakes are high, but the rewards could be transformative.

Chapter 6

The Mechanics of a Universal Voting System

A universal voting system leveraging biometric identification, thumbprints, and facial recognition offers a glimpse into the future of seamless, secure, and accessible democratic participation. This chapter explores the practical mechanics of such a system: how it would work, who would oversee its operations, potential challenges for certain populations, and measures to incentivize participation.

The Voting Process: How Would It Work?

The universal voting system would rely on cutting-edge technology to ensure security, accuracy, and ease of use. Here's a step-by-step breakdown of the process:

1 Biometric Registration:

○ Before using the system, individuals would register their biometric data—thumbprints, facial recognition, and possibly retinal scans—alongside other personal information, such as government-issued ID numbers.

○ This data would be securely stored in a decentralized, encrypted system to prevent breaches.

2 Voting Locations and Accessibility:

○ Voters could cast their ballots at designated polling stations equipped with biometric scanners and touchscreens.

○ Alternatively, secure voting kiosks could be placed in public locations like libraries, malls, and government buildings.

○ Remote voting would also be possible through a secure mobile app or website, provided the device has the necessary biometric scanning capabilities.

3 Casting Your Vote:

○ At a polling station, voters would approach a voting machine, which would prompt them to scan their thumbprint or face.

○ The system would verify their identity against the encrypted database and display their voter profile.

○ Once verified, voters could proceed to the ballot screen, selecting their choices on a user-friendly interface.

○ After reviewing their selections, voters would confirm their choices, and the system would securely record the vote.

4 Accessibility Features:

○ For individuals with disabilities or those unable to use thumbprints, alternative methods such as facial recognition or voice authentication could be employed.

○ The interface would include options for larger text, audio prompts, and multiple languages to ensure inclusivity.

Oversight and Management: Who Controls the System?

To maintain trust in the system, control and oversight would need to be carefully balanced:

- **Independent Governance**: An independent, non-partisan body would oversee the software and hardware, ensuring transparency and accountability.

- **Public-Private Collaboration**: While private companies might develop and maintain the software, strict government oversight and third-party audits would prevent conflicts of interest or tampering.

- **Open-Source Technology**: Making the voting software open-source could allow for public scrutiny, enhancing trust and reducing the risk of fraud.

- **Cybersecurity Protocols**: Regular security audits, blockchain technology for vote recording, and robust encryption would safeguard the system from hacking or manipulation.

Potential Challenges and Solutions

Challenges for Certain Populations

1 Physical Limitations:

○ Some individuals, such as those with missing fingers or facial differences, may face difficulties using thumbprint or facial recognition.

o **Solution**: Offering multiple identification methods (e.g., retinal scans, voice recognition) ensures that no one is excluded.

2 Technological Familiarity:

○ Older voters or those unfamiliar with digital technology might feel intimidated by the system.

o **Solution**: Providing on-site assistance and offering practice sessions could help these voters gain confidence.

3 Distrust of Technology:

○ Some individuals may distrust biometric systems due to privacy concerns or fear of surveillance.

o **Solution**: Transparency in how data is stored and used, along with stringent privacy protections, could alleviate these fears.

Addressing Objections to the System

- **Preference for Manual Ballots**:

○ Some voters might prefer traditional paper ballots due to familiarity or perceived reliability.

○ **Solution**: Offering a hybrid system where voters can choose between digital or paper ballots could ease the transition.

- **Resistance to Change**:

○ People resistant to adopting new methods might need reassurance.

○ **Solution**: Educational campaigns highlighting the system's ease of use and benefits could address these concerns.

The Role of Election Workers

While the system would automate many aspects of the voting process, human oversight would still be necessary:

- **On-Site Assistance**: Poll workers would be available to guide voters through the process and address technical issues.

- **System Monitoring**: Trained technicians would monitor the machines to ensure they function correctly.

- **Voter Education**: Workers could also help educate voters about the system, ensuring everyone can participate.

Remote Voting: Voting from Home

A universal voting system could allow citizens to vote from the comfort of their homes:

- **Biometric Authentication**: Voters would use their smartphones or computers equipped with biometric scanners to verify their identity.
- **Secure Connection**: Votes would be transmitted through encrypted channels to ensure security.
- **Accessibility**: This option would be especially beneficial for individuals with mobility issues or those living in remote areas.

Incentivizing Participation

To encourage widespread use of the system, governments could offer incentives for voting:

- **Access to Educational Materials**: Voters could receive free resources about the issues and candidates on the ballot, promoting informed decision-making.
- **Tax Credits or Discounts**: Small tax credits, public transit discounts, or other benefits could reward civic participation.
- **Recognition Programs**: Public acknowledgment, such as certificates or badges, could motivate engagement.

Conclusion

A universal voting system using biometric authentication promises a more accessible, secure, and efficient democratic process. While challenges exist, thoughtful design, robust oversight, and comprehensive education can address potential issues. By streamlining the act of voting and incentivizing participation, this system could pave the way for a more engaged and equitable society

Chapter 7

How to Integrate Ballots into a Universal Voting System

Creating a universal voting system that allows anyone to vote on anything, anytime, and anywhere requires a seamless process for introducing ballots into the system. This chapter explores how ballots can be added to the system, how people could be incentivized to contribute, and the broader implications of a user-friendly and participatory design for policymaking and governance.

Introducing Ballots into the System

For a universal voting system to succeed, the process of adding ballots must be simple, transparent, and inclusive. Here's how it could work:

1 Community Proposals:

○ Citizens, organizations, or government officials could propose ballot measures through a centralized platform.

○ Proposals would require a basic outline, including the issue at hand, potential solutions, and any associated costs or benefits.

2 Validation and Review:

○ Proposed ballots would undergo a validation process to ensure they meet legal and procedural requirements.

○ A neutral oversight body could review proposals for clarity, fairness, and feasibility before approving them for public voting.

3 Transparency and Communication:

○ Once approved, ballots would be publicly listed with detailed information, including context, potential impacts, and arguments for and against the measure.

○ Educational resources, such as videos, infographics, and expert analyses, could be provided to help voters make informed decisions.

4 Customizing Ballot Access:

○ Ballots could be localized based on the voter's address, ensuring individuals only vote on issues relevant to their community.

○ National or global issues could be available to all eligible voters, creating a tiered system for different levels of governance.

Incentivizing Participation in Ballot Creation

One of the key challenges is encouraging individuals and groups to create and propose ballots. Drawing inspiration from platforms like Facebook, which thrives on user-generated content, the following strategies could be employed:

1 Gamification:

○ Introduce rewards for submitting well-received or widely-supported ballot proposals, such as badges, recognition on leaderboards, or small monetary incentives.

○ Offer milestones for frequent contributors, encouraging sustained engagement.

2 Simplified Submission Tools:

○ Create intuitive interfaces for proposing ballots, such as drag-and-drop templates or guided forms, reducing the technical barriers to participation.

○ Allow for collaborative proposals, where multiple users can work together to refine and submit ideas.

3 Community Recognition:

○ Highlight successful proposals and their contributors, offering public acknowledgment through social media, newsletters, or community events.

○ Provide opportunities for contributors to discuss their proposals with policymakers or present them in public forums.

4 Educational Campaigns:

○ Run outreach programs to educate individuals about the importance of proposing ballots and how their input can directly influence policies.

○ Offer workshops, webinars, and online tutorials to demystify the process and encourage broader participation.

Legitimizing Policy Through Voting Results

A universal voting system not only empowers individuals but also provides legitimacy to policymaking by grounding decisions in public consensus. Here's how:

1 Empowering Policymakers:

○ Policymakers can use voting results as a clear mandate to enact decisions, reducing opposition and building public trust.

○ For contentious issues, voting outcomes can serve as a democratic benchmark, guiding leaders toward widely supported solutions.

2 Encouraging Participatory Governance:

○ By allowing citizens to vote on specific policies, the system fosters a sense of ownership and accountability.

○ Participation in decision-making strengthens the relationship between the government and its constituents, reducing apathy and cynicism.

3 Local and Personalized Decision-Making:

○ Communities could vote on hyper-local issues, such as:

■ Whether streetlights should remain on all night or operate on a timed schedule.

■ The timing and frequency of garbage collection.

■ Allocating funds for community projects, like repurposing unused pavement or installing green spaces.

○ These granular decisions allow policies to reflect the unique needs and preferences of each community.

Making the System Easy to Use

For the system to thrive, ease of use is paramount. Here are some features that could streamline the user experience:

1 Mobile and Desktop Accessibility:

○ The voting platform would be accessible via smartphones, tablets, and computers, ensuring broad reach and convenience.

2 User-Friendly Interfaces:

○ Intuitive design with clear instructions, simple navigation, and accessible options for individuals with disabilities.

○ Multi-language support to accommodate diverse populations.

3 Real-Time Feedback:

○ Voters could see immediate confirmations of their submissions and track the progress of ballot proposals and voting results.

4 Integrated Education Tools:

○ Links to resources, explainer videos, and FAQs directly within the platform would help voters understand the issues and make informed decisions.

Potential Incentives for Voting

To further encourage participation, governments could offer tangible benefits for voting:

1 Educational Resources:

○ Free access to educational materials related to the issues on the ballot, fostering informed citizenship.

2 Tax Credits or Financial Rewards:

○ Small tax credits, utility bill discounts, or public transportation vouchers could incentivize voter turnout.

3 Community Rewards:

○ Communities with high voter participation rates could receive additional funding for local projects or public services.

4 Civic Recognition:

○ Public acknowledgment of high participation rates or individual contributions could motivate engagement.

Conclusion

Integrating ballots into a universal voting system requires thoughtful design and widespread incentives to encourage participation. By making the process user-friendly and providing legitimacy to policymaking, this system has the potential to revolutionize governance. Whether deciding on local matters like streetlight schedules or national policies on healthcare, the universal voting system empowers individuals to shape their world, one vote at a time.

Chapter 8

Maintaining the Integrity of a Universal Voting System

Public trust is the cornerstone of any voting system, and this is especially true for a universal voting system reliant on technology. While such a system promises accessibility, efficiency, and inclusivity, it also raises concerns about accuracy, tampering, and corruption. This chapter explores how to maintain the integrity of the system, address public skepticism, and ensure that every vote counts.

Addressing Public Concerns: Ensuring Accuracy and Security

1 System Transparency:

○ The system's code and operation processes should be open-source and publicly auditable. This transparency allows independent experts to scrutinize and verify its integrity.

○ Clear, publicly available documentation on how

votes are recorded, transmitted, and tallied would help build trust.

2 Multi-Layered Security:

- **Encryption**: All data, from biometric authentication to vote transmission, must be encrypted with state-of-the-art technology.

- **Blockchain Technology**: Utilizing blockchain for vote recording could create an immutable, transparent ledger, allowing real-time verification while maintaining voter anonymity.

- **Redundant Systems**: Multiple backup servers and data replication would protect against system failures or data loss.

3 Independent Oversight:

- A non-partisan, independent electoral commission would oversee the system, ensuring fairness and preventing conflicts of interest.

- Regular third-party audits by cybersecurity firms and international election observers would verify the system's integrity.

4 Tamper-Proof Hardware:

- Voting machines and servers would be equipped with tamper-detection mechanisms that alert authorities to any unauthorized access attempts.

- Secure access protocols, such as biometric

authentication for election workers, would restrict system access to authorized personnel only.

Verification Mechanisms: Ensuring Votes Are Counted

1 Voter-Verified Audit Trails (VVAT):

○ After casting a vote, each voter would receive a printed receipt with a unique code (not tied to their identity) that reflects their choices.

○ Voters could use this code to confirm their vote was recorded accurately via a secure, online portal.

2 Post-Election Audits:

○ Randomized manual audits would compare the digital records with printed backups to ensure accuracy.

○ These audits would be conducted in public view, with representatives from all major parties and independent observers present.

3 Real-Time Reporting:

○ The system could provide real-time vote tally updates visible to the public while protecting individual voter anonymity.

4 Public Feedback Loop:

○ A feedback system would allow voters to report any anomalies they experience during the process, triggering an immediate investigation.

Oversight and Accountability

1 Selection of Oversight Bodies:

○ Members of the electoral commission would be chosen through a transparent, merit-based process, involving input from multiple stakeholders, including civil society organizations and international experts.

○ Term limits and regular performance reviews would prevent entrenchment and ensure accountability.

2 Monitoring Oversight Performance:

○ The commission's activities would be subject to regular public reporting and evaluation.

○ Independent citizen panels or watchdog organizations could review their performance and recommend improvements.

3 Public Access to Information:

○ Detailed reports on the system's performance, security updates, and audit results would be made publicly available to ensure transparency.

Indicators of System Success

1 Quantitative Verification:

○ High voter turnout rates and minimal error margins would indicate a functional system.

○ Comprehensive audits with consistent results would reinforce confidence.

2 Qualitative Feedback:

○ Positive public sentiment, as reflected in surveys and media reports, would demonstrate trust in the system.

○ Reduced complaints about voting issues would signify improvements.

3 Multiple Verification Layers:

○ Cross-referencing independent audits, blockchain records, and public feedback would create a robust verification framework.

Red Flags: Identifying and Addressing Issues

1 Irregularities in Vote Tallies:

○ Sudden, unexplained spikes in votes or discrepancies between digital and printed records should prompt immediate investigations.

2 System Downtime or Failures:

○ Frequent outages or slowdowns could indicate vulnerabilities or potential tampering.

3 Reports of Voter Disenfranchisement:

○ Complaints of voters being unable to access the system or verify their votes must be addressed promptly.

4 Security Breaches:

○ Any unauthorized access attempts or breaches should trigger transparency measures, including public reporting and third-party investigations.

Addressing Frequently Asked Questions (FAQs)

1 How do I know my vote counts?

○ You will receive a receipt with a unique code to verify your vote was recorded accurately.

2 What if the system gets hacked?

○ Multi-layered security measures, including encryption and blockchain, protect against hacking. Regular audits ensure any discrepancies are identified and resolved.

3 Who ensures the system is fair?

○ An independent electoral commission, subject to public scrutiny and third-party audits, oversees the system.

4 Can my vote be traced back to me?

○ No. The system is designed to ensure anonymity while allowing voters to verify their participation.

Conclusion

Maintaining the integrity of a universal voting system requires a combination of transparency, security, and robust oversight. By addressing public concerns, implementing multiple verification mechanisms, and ensuring accountability, such a system can build trust and confidence in the democratic process. The ultimate proof of its success will be a well-informed, engaged electorate and decisions that reflect the collective will of the people.

Chapter 9

The Technical Challenges of Setting Up and Maintaining a Universal Voting System

Establishing a universal voting system that allows anyone to vote on anything, anywhere, at any time is a monumental technical challenge. From the development of secure and accessible software to the deployment of voting kiosks in remote areas, the system must overcome numerous hurdles to ensure inclusivity, reliability, and scalability. This chapter delves into the technical requirements for setting up, operating, and maintaining such a system, exploring the technology needed, the implementation process, and solutions for challenges in diverse environments.

Technology Requirements

1 Core Software Development:

○ **Open-Source Voting Software**: To ensure transparency and security, the software must be open-source and publicly auditable.

o **Compatibility**: The software should be compatible with various devices, including smartphones, tablets, desktops, and voting kiosks.

o **Biometric Integration**: The system must support biometric authentication, such as fingerprint scanning, facial recognition, and potentially voice recognition.

o **User-Friendly Interfaces**: Simple and intuitive design is essential for accessibility, especially for individuals with limited technological experience.

o **Localization**: Multi-language support and cultural adaptations are critical to serving diverse populations.

2 Hardware Requirements:

o **Secure Servers**: High-performance servers with robust encryption to handle data storage, processing, and transmission securely.

o **Voting Kiosks**: Rugged and tamper-proof kiosks for deployment in public spaces and remote areas, equipped with:

- Touchscreens.
- Biometric scanners.
- Printers for voter-verified paper trails.

o **Mobile and Desktop Compatibility**: Ensure the system works seamlessly across personal devices with minimal hardware requirements.

3 Infrastructure Needs:

o **Reliable Internet Connectivity**: High-speed, secure internet access is essential for real-time vote transmission and system updates.

o **Offline Voting Capabilities**: Develop an offline mode for areas with limited connectivity, with data synchronization once a connection is established.

o **Energy Solutions**: Solar-powered kiosks or battery backups for regions with unstable power supplies.

Implementation Challenges

1 Software Deployment:

○ Can the software be downloaded and installed on personal devices? Yes, but safeguards must ensure that only authorized versions are used to prevent tampering.

○ Pre-installing the software on all new devices could be mandated by governments or encouraged through partnerships with manufacturers.

2 Kiosk Deployment in Underserved Areas:

o **Site Selection**: Identify accessible locations for kiosks in rural and remote areas, such as schools, community centers, and post offices.

o **Local Partnerships**: Collaborate with local organizations to install and maintain kiosks.

o **Accessibility Features**:

■ Braille interfaces and audio assistance for visually impaired users.

■ Simplified navigation for users with limited literacy.

■ Physical accessibility for individuals with mobility challenges.

3 Data Security and Privacy:

o **Encryption**: End-to-end encryption for all data to prevent unauthorized access.

o **Multi-Factor Authentication**: Combine biometrics with other methods, such as PINs or QR codes, for enhanced security.

o **Decentralized Data Storage**: Use distributed ledger technology (e.g., blockchain) to prevent single points of failure.

4 Operational Challenges:

o **System Scalability**: Design the system to handle millions or even billions of votes without delays or crashes.

o **Real-Time Monitoring**: Implement tools to detect and respond to technical issues or security threats in real-time.

o **Regular Updates**: Ensure the system can be updated without disrupting operations or risking data integrity.

Maintenance and Continuous Improvement

1 Technical Support:

○ Establish 24/7 technical support teams to address issues promptly.

○ Train local technicians to maintain and repair kiosks in their communities.

2 System Testing and Audits:

○ Conduct regular stress tests and simulations to identify vulnerabilities.

○ Perform frequent security audits by independent experts.

3 User Feedback and Adaptation:

○ Collect feedback from users to identify pain points and areas for improvement.

○ Continuously adapt the system to meet evolving technological and social needs.

Expanding Accessibility

1 Designing for Universal Access:

○ Develop interfaces that are easy to navigate regardless of a user's education level or technical experience.

○ Incorporate audio guidance and visual aids to assist individuals with disabilities.

2 Community Outreach and Education:

○ Launch educational campaigns to teach citizens how to use the system.

○ Provide hands-on demonstrations and support, especially in areas with low digital literacy.

3 Incentives for Participation:

○ Encourage early adoption by offering incentives, such as tax credits, discounts on utilities, or access to educational resources.

Ensuring Maximum Participation

The ultimate goal is to design a system that enables the largest number of people to vote as frequently as needed. Key strategies include:

1 Broad Deployment:

○ Install kiosks in urban and rural areas to ensure no one is left out.

○ Provide portable devices or voting kits for individuals in extremely remote locations.

2 Flexible Voting Options:

○ Allow voting through personal devices for those with internet access.

○ Offer extended voting windows to accommodate varying schedules.

3 Public Engagement:

○ Promote the system as a civic duty and a tool for empowerment.

○ Highlight success stories of how voting influences local and national policies.

Conclusion

The technical challenges of a universal voting system are significant but not insurmountable. By leveraging cutting-edge technology, ensuring robust infrastructure, and prioritizing accessibility, the system can empower individuals to participate in the democratic process like never before. With the right investments in development, deployment, and maintenance, a universal voting system can be a transformative force for inclusive governance.

Chapter 10

Overcoming Political Opposition to a Universal Voting System

The introduction of a universal voting system would undoubtedly face resistance, particularly from established political parties. Democrats and Republicans, the dominant political forces in the United States, might oppose such a proposal for different reasons. However, the potential benefits of such a system—eliminating gerrymandering, increasing voter participation, and ensuring more equitable representation—could ultimately gain bipartisan support. This chapter explores the challenges, strategies to address opposition, and steps to educate the public and implement the system.

Resistance from Dominant Political Parties

1 Democratic Concerns:

○ Democrats often advocate for expanded voting

rights, but they might fear losing control in areas where they traditionally hold a majority.

○ Concerns about technical limitations or vulnerabilities in the system could be leveraged to justify opposition.

○ Skepticism about whether the system would address systemic issues like voter suppression might also arise.

2 Republican Concerns:

○ Republicans might oppose the system due to fears of increased voter turnout, which they sometimes perceive as unfavorable to their electoral chances.

○ Concerns about voter fraud, even if unfounded, might be used to justify resistance.

○ Opposition to eliminating gerrymandering, which often benefits Republicans, could also fuel resistance.

3 Shared Concerns:

○ Both parties might resist ceding control over redistricting and other mechanisms that give them electoral advantages.

○ Questions about who controls and oversees the system could lead to bipartisan hesitation.

Strategies to Build Political Support

1 Framing the System as Non-Partisan:

○ Emphasize that the system benefits all citizens and ensures fair representation.

○ Highlight safeguards that prevent any party from manipulating the system for partisan gain.

2 Appealing to Party Values:

○ For Democrats, emphasize inclusivity, voter participation, and the system's potential to combat voter suppression.

○ For Republicans, stress the system's efficiency, security, and ability to eliminate voter fraud.

3 Bipartisan Commission:

○ Establish a bipartisan commission to design and oversee the system, ensuring that neither party feels excluded.

○ Include independent experts and representatives from smaller parties to broaden support.

Educating the Public

1 Public Awareness Campaigns:

○ Use social media, television, and community events to explain the benefits and functionality of the system.

○ Create accessible educational materials, including videos, infographics, and workshops, to reach diverse audiences.

2 Highlighting Success Stories:

○ Showcase examples of similar systems implemented successfully in other countries or regions.

○ Provide data on how universal voting systems increase participation and improve representation.

3 Addressing Misconceptions:

○ Counter misinformation about voter fraud, technical vulnerabilities, and potential biases in the system.

○ Engage trusted community leaders to advocate for the system and address local concerns.

Implementation Timeline

1 Initial Steps (1-2 Years):

○ Form a bipartisan commission to study feasibility and design the system.

○ Pilot the system in a few municipalities or states to identify and address challenges.

2 Mid-Term Steps (3-5 Years):

○ Expand the system to include more states, with ongoing monitoring and adjustments.

○ Develop and distribute educational materials to ensure widespread understanding and acceptance.

3 Full Implementation (5-10 Years):

○ Roll out the system nationwide, ensuring that all

infrastructure, training, and oversight mechanisms are in place.

○ Continuously improve the system based on user feedback and technological advancements.

Potential Changes to Voting Issues

1 Elimination of Gerrymandering:

○ Universal voting would make gerrymandering obsolete, as district boundaries would no longer determine representation.

○ This change would result in more competitive elections and fairer outcomes.

2 Increased Voter Participation:

○ By making voting accessible anytime, anywhere, participation rates would likely rise significantly.

○ Groups traditionally underrepresented in voting, such as low-income individuals and people with disabilities, would benefit.

3 Simplified Election Processes:

○ Universal voting could reduce the need for costly and complex election logistics, such as maintaining polling places and printing ballots.

Limitations and Challenges

1 Technical Limitations:

○ Ensuring the system's security and reliability requires significant investment and expertise.

○ Accessibility for individuals without internet access or technological proficiency remains a challenge.

2 Resistance to Change:

○ Some voters may distrust the system or prefer traditional voting methods.

○ Political resistance from dominant parties could delay implementation.

3 Equity Concerns:

○ Ensuring equal access to the system for all communities, particularly marginalized groups, requires targeted efforts and resources.

Steps to Enact the System

1 Legislative Action:

○ Pass federal legislation mandating the adoption of the universal voting system.

○ Allocate funding for development, deployment, and maintenance.

2 Public Referendums:

○ Allow citizens to vote directly on adopting the system, building grassroots support and legitimacy.

3 International Collaboration:

○ Learn from countries with advanced voting systems to adopt best practices and avoid common pitfalls.

Conclusion

A universal voting system has the potential to transform democracy by making it more inclusive, equitable, and efficient. While political resistance and technical challenges are significant, they are not insurmountable. By addressing concerns, educating the public, and building bipartisan support, the system can be implemented successfully, ultimately empowering citizens and strengthening democratic governance.

Chapter 11

The Economic Costs of a Universal Voting System Powered by Artificial Intelligence

The implementation of a universal voting system utilizing artificial intelligence (AI) would inevitably involve significant costs, particularly during the initial development and deployment stages. However, over time, such a system has the potential to reduce the overall costs associated with elections at the local, state, and national levels. This chapter will compare the current costs of running elections in various U.S. cities and states with the projected costs of an AI-powered universal voting system, highlighting potential savings and trade-offs.

Current Costs of Elections

The cost of running elections in the United States varies widely depending on the scale and location. Below are examples of the financial burdens of conducting elections in New York City, Minneapolis, and Portland:

1 New York City:

○ In the 2020 presidential election, New York City spent approximately $120 million on election-related activities, including staffing, polling station setup, and equipment maintenance.

○ Costs were driven by the high number of voters (over 4.5 million registered voters) and the need to maintain hundreds of polling locations.

2 Minneapolis:

○ Minneapolis, with a smaller population, spent about $2.5 million on its 2021 municipal elections, which included mayoral and city council races.

○ Costs included paper ballots, voting machine rentals, and poll worker salaries.

3 Portland:

○ Portland's mail-in voting system, used in the 2022 elections, cost the city approximately $3 million.

○ While mail-in voting reduces the need for physical polling locations, the costs of printing, mailing, and processing ballots remain significant.

At the national level, the cost of the 2020 U.S. presidential election exceeded $14 billion, when including campaign expenditures. Direct government spending on election administration was a smaller portion of this total but still substantial, reaching hundreds of millions of dollars across states.

Costs of an AI-Powered Universal Voting System

An AI-powered universal voting system would involve distinct cost categories:

1 Upfront Costs:

o **Development**: Designing and programming the software for secure voting and identity verification could cost an estimated $2-5 billion, depending on the system's complexity and scope.

o **Hardware**: Deploying biometric kiosks and secure servers nationwide could require an additional $10 billion.

o **Training**: Educating election workers and technicians to maintain and operate the system could cost approximately $500 million.

2 Ongoing Costs:

o **Maintenance**: Regular updates to the software, cybersecurity measures, and hardware repairs would cost an estimated $500 million annually.

o **User Support**: Providing technical support and accessibility assistance for voters could require an annual budget of $300 million.

o **Data Storage**: Storing encrypted voting records securely in the cloud or on dedicated servers would add another $200 million annually.

3 Savings:

○ By eliminating paper ballots, physical polling locations, and temporary staffing, the system could save billions of dollars annually.

○ For example, the U.S. spends around $1 billion on paper ballots and associated logistics each election cycle, costs that would be largely eliminated.

Comparative Analysis

1 Short-Term Costs:

○ The initial costs of implementing a universal AI voting system would be significantly higher than a single election cycle under the current system. For example, the $12-15 billion required for development and deployment would far exceed the $120 million spent on New York City's 2020 elections.

2 Long-Term Savings:

○ Over time, the universal system would generate savings by reducing recurring expenses such as poll worker salaries, ballot printing, and polling station setup. Assuming an average election cycle cost of $3 billion nationwide, the system could pay for itself within 5-6 election cycles.

3 Hidden Costs and Benefits:

○ The system's efficiency and accuracy could reduce the costs of litigation and recounts, which have become common in contentious elections.

○ Increased voter participation, while a social benefit, could introduce additional costs for data storage and system capacity.

Potential Economic Impacts

1 For Governments:

○ The savings generated by an AI-powered system could be reallocated to other public services, such as education and healthcare.

○ However, the upfront costs might require bond issuance or tax increases, which could face political resistance.

2 For Private Sector Vendors:

○ Companies specializing in cybersecurity, AI development, and biometric hardware would benefit economically from the system's implementation.

○ Conversely, industries reliant on traditional election logistics, such as ballot printing companies, might face economic challenges.

3 For Voters:

○ A universal voting system could eliminate indirect costs for voters, such as taking time off work to vote or traveling to polling locations.

Limitations and Risks

1 Equity Concerns:

○ Rural and low-income areas might require additional investments to ensure equal access to the system, increasing costs in the short term.

2 Technological Risks:

○ Cybersecurity threats, such as hacking or data breaches, could undermine trust in the system and necessitate costly countermeasures.

3 Public Resistance:

○ Resistance to the system's high initial costs could delay implementation, increasing overall expenses.

Conclusion

While the upfront costs of an AI-powered universal voting system would be substantial, the long-term savings and efficiencies could outweigh the initial investment. By eliminating many of the recurring expenses associated with traditional elections, the system has the potential to save billions of dollars over time. However, careful planning, equitable implementation, and transparent communication with the public will be essential to ensure the system's success and sustainability.

Chapter 12

Universal Voting Systems: A Labor and Fairness Perspective

The idea of implementing a universal voting system has gained traction as an avenue to ensure equitable access to the democratic process. But beyond its ethical and societal implications, how does such a system compare to our current patchwork voting framework when it comes to labor, cost, fairness, and economic efficiency? This chapter examines these questions, offering an analysis of the labor dynamics, benefits, and challenges associated with a universal voting system.

Labor Requirements: Current System vs. Universal Voting System

Under the current voting system, the amount of labor required varies significantly by state and county due to differences in election administration. States operate with distinct rules on voter registration, mail-in voting, early voting, and Election Day operations, creating inefficiencies and redundancies. Election officials, poll workers, and temporary staff are often overworked, especially during major election cycles. Additionally, voter rolls must be continually maintained, leading to fragmented and labor-intensive processes.

A universal voting system, by contrast, would centralize and standardize election procedures. Automated voter registration, real-time updates to voter rolls, and universal mail-in ballots or digital voting platforms would dramatically reduce the administrative burden. Fewer temporary workers would be required for manual tasks like voter registration or polling station setup, as technology would streamline many of these processes. However, this shift would necessitate a skilled workforce adept at maintaining and securing digital systems, as well as operating centralized databases. Overall, the total labor required may decrease, but the nature of the work would change, emphasizing technical expertise over manual administration.

Labor Benefits and Displacement

For workers in the election administration field, a universal voting system could offer numerous benefits, including:

- **Job Stability**: By reducing seasonal spikes in demand for temporary poll workers, universal systems could provide year-round roles for election staff focused on system maintenance, voter education, and cybersecurity.

- **Specialization Opportunities**: Roles in data management, IT support, and system development would increase, providing skilled employment opportunities.

- **Reduced Stress**: Automating time-intensive tasks like voter roll updates and ballot counting would alleviate pressure on overworked election officials.

However, this transition would also displace certain workers. Temporary poll workers, who often rely on short-term income during election seasons, could find fewer opportunities. Similarly, roles focused on manual tasks, such as ballot distribution or voter registration drives, would likely diminish. Transitioning these workers into technical roles would require retraining initiatives and government support.

Fairness of a Universal Voting System

A universal voting system's primary advantage lies in its potential to enhance fairness. Standardizing processes across jurisdictions eliminates disparities that currently disenfranchise certain populations, such as those in rural areas or states with restrictive voting laws. Features like same-day registration, universally accessible voting options (e.g., mail-in, online, or in-person), and multilingual ballot access would ensure equal participation opportunities for all voters.

However, concerns about fairness could emerge if the system's implementation creates barriers for specific groups. For example, reliance on digital platforms might disadvantage individuals without reliable internet access or digital literacy. Addressing these disparities would require significant investment in public infrastructure and education.

Cost Analysis

The upfront costs of implementing a universal voting system could be substantial, encompassing investments in technology, cybersecurity, and infrastructure. These expenses include:

- Developing a centralized voting platform
- Training election staff in new systems
- Implementing robust security measures to protect against cyber threats

Over time, however, the system would likely reduce costs by eliminating redundancies and inefficiencies. Fewer temporary workers, streamlined voter roll maintenance, and reduced reliance on paper ballots would lower ongoing expenses. Additionally, increased voter participation—a likely outcome of a universal system—could strengthen democratic institutions, yielding long-term societal benefits that outweigh initial costs.

The Broader Implications for Workers

Workers in industries adjacent to the voting system, such as printing and mailing services, might experience shifts in demand. For example, universal digital voting would reduce the need for printed ballots and mailing services, potentially impacting jobs in these sectors. Conversely, increased reliance on technology would spur growth in cybersecurity and software development fields, creating new employment opportunities.

Conclusion: A Path Forward

Adopting a universal voting system offers clear advantages in terms of fairness, efficiency, and labor stability, but it also poses challenges for certain workers and industries. Transitioning to such a system would require thoughtful planning, significant investment in infrastructure and training, and a commitment to addressing potential disparities.

Ultimately, a universal voting system has the potential to create a more equitable and streamlined democratic process. By reallocating labor from manual and temporary roles to skilled, year-round positions, it could provide stability for workers while enhancing the integrity and accessibility of elections. As we consider this transition, policymakers must weigh the benefits of increased fairness and reduced inefficiency against the costs and disruptions that such a monumental shift would entail.

Chapter 13

Integrating a Universal Voting System into AI-Driven Cities

Integrating a Universal Voting System into AI-Driven Cities

As cities increasingly adopt artificial intelligence (AI) and robotic technologies to manage essential services, integrating a universal voting system into this advanced infrastructure offers groundbreaking potential. This chapter explores how such a system could function within an AI-powered city where sanitation, information dissemination, and other services are automated. By leveraging AI sensors, robotics, and other smart technologies, universal voting systems can create a direct, participatory link between civic engagement and service optimization, empowering residents to shape their urban environment in real-time.

AI and Robotics in Urban Services

AI and robotics already play a critical role in modern urban management. Autonomous systems handle sanitation tasks, AI platforms streamline public information campaigns, and citywide sensors monitor everything from traffic to air quality. These technologies could be seamlessly integrated into a universal voting system to make governance more responsive and participatory.

For example, residents could vote on:

- **Sanitation Services**: Households could decide how frequently trash is collected. Votes would be securely cast via a centralized platform, and AI algorithms would compile the data. Autonomous trash collection robots would adjust their schedules and routes accordingly, ensuring that service matches local demand.

- **Crime Prevention**: Citizens could prioritize investments in public safety initiatives, such as installing more surveillance cameras, deploying additional security drones, or funding community programs. AI-driven policing tools could then reallocate resources based on these priorities.

- **Healthcare Access**: Voters might choose to direct funding toward mobile clinics, telehealth services, or AI-powered diagnostics in underserved areas. The system would analyze voting outcomes to adjust healthcare resources dynamically.

- **Housing and Education**: Residents could vote on zoning laws, affordable housing projects, or the introduction of new school programs. AI systems would use these results to guide urban planning and resource distribution.

- **Universal Basic Income (UBI)**: Citizens could vote on the allocation of funds for UBI programs, tailoring financial support to community needs. AI algorithms could ensure that payments are distributed efficiently and equitably based on voting outcomes.

Leveraging AI Sensors and Smart Systems

AI sensors and smart equipment form the backbone of an integrated voting and service-delivery system. These technologies would collect, process, and act upon voter preferences, creating a real-time feedback loop between citizens and city infrastructure. Key components include:

- **Smart Sensors**: Sensors embedded in city infrastructure could monitor conditions like waste levels, traffic patterns, or energy consumption. This data could inform voters about current challenges and help them make more informed decisions.

- **Dynamic Interfaces**: AI-driven platforms could provide personalized voting interfaces, highlighting issues relevant to each voter's household or neighborhood. For example, a household in an area with high pollution levels might receive prompts to vote on environmental initiatives.

- **Autonomous Implementation**: Once voting results are finalized, AI systems could autonomously execute decisions. For instance, if residents vote to increase school funding, robotic construction crews might begin work on new facilities, or AI learning platforms could be deployed in classrooms.

- **Resource Allocation**: AI systems could dynamically adjust city budgets and resource distribution based on voting outcomes, ensuring that services like sanitation, public safety, and healthcare are aligned with community priorities.

Enhancing Civic Engagement Through AI

Integrating a universal voting system into AI-powered cities has the potential to increase civic engagement by simplifying the voting process and making it more relevant to everyday life. Features like real-time updates on the impact of voter decisions and gamified participation incentives could encourage broader involvement. For example:

- **Impact Dashboards**: Residents could view dashboards showing how their votes directly influence city operations, such as improved trash collection schedules, reduced crime rates, or enhanced healthcare access.

- **Gamification**: AI systems could reward active participation with virtual badges or discounts on city services, fostering a sense of community and accomplishment.

Addressing Security and Fairness

Ensuring the security and fairness of an AI-integrated voting system is paramount. Cities must implement measures to prevent fraud, protect voter privacy, and ensure equitable access. Key considerations include:

- **Blockchain Technology**: Secure, tamper-proof voting records could be maintained using blockchain systems.

- **Biometric Authentication**: Facial recognition or fingerprint scans could ensure that only eligible residents participate.

- **Digital Inclusivity**: To prevent disenfranchisement, cities must offer alternative voting methods for residents without internet access or digital literacy.

Challenges and Opportunities

While the integration of universal voting systems with AI offers immense potential, it also presents challenges:

- **Privacy Concerns**: Striking a balance between data collection and individual privacy rights is essential. Transparent policies and opt-out options must be provided.

- **Cost**: Developing and maintaining advanced AI-driven voting infrastructure requires significant investment. However, these costs could be offset by long-term efficiencies and increased voter engagement.

- **Adaptation**: Residents may require education on how to use the new system effectively, necessitating public outreach campaigns.

Conclusion: Transforming Democracy Through Technology

Integrating a universal voting system into AI-driven cities represents a transformative step toward a smarter, more inclusive democracy. By empowering residents to directly influence decisions on sanitation, crime prevention, healthcare, housing, education, and universal basic income, this approach creates a responsive urban environment tailored to the needs of its citizens. While challenges remain, the potential benefits of enhanced engagement, efficiency, and fairness make this vision a compelling path forward for modern governance.

Chapter 14

Universal Voting Systems in Auto-Free Cities

As cities transition to being car-free or significantly reducing automobile dependency, the role and focus of a universal voting system shift. Without the need to regulate traffic flow, parking, emissions, and other car-related issues, the voting system could instead emphasize other aspects of urban living. This chapter explores the benefits and consequences of implementing a universal voting system in an auto-free city and highlights how governance priorities evolve in such environments.

Benefits of a Universal Voting System in an Auto-Free City

1 Streamlined Urban Planning:

○ Without cars, cities can repurpose streets and parking lots into green spaces, pedestrian zones, and public infrastructure. A universal voting system allows residents to have a direct say in how these spaces are developed, fostering community-driven urban planning.

○ Citizens could vote on whether to prioritize bike lanes, public gardens, or additional housing developments, ensuring that the city evolves to meet the needs of its residents.

2 Enhanced Focus on Public Transport:

○ Public transportation becomes the backbone of an auto-free city. Voters could influence decisions on routes, frequency, and investments in sustainable transit solutions like trams, subways, and e-buses.

○ Real-time feedback from residents through the voting system could help optimize transit schedules and infrastructure.

3 Improved Air Quality and Health:

○ With fewer or no cars, cities experience better air quality and reduced noise pollution. Universal voting systems could enable residents to prioritize health initiatives, such as planting more trees, expanding parks, or funding wellness programs.

4 Reallocation of Resources:

○ Funds previously spent on maintaining roads, traffic management, and parking infrastructure can be redirected to other areas like education, housing, or renewable energy. Voters could play a key role in deciding how these resources are reallocated.

5 Increased Civic Engagement:

○ The absence of car-related issues allows voters to focus on broader societal needs, such as public safety, cultural programs, and social equity. This shift may encourage higher levels of civic participation as residents feel more connected to these issues.

Consequences and Challenges of Universal Voting in an Auto-Free City

1 Reduced Scope of Governance Topics:

○ The elimination of car-related concerns may simplify urban governance, but it could also reduce the range of topics on which residents feel compelled to vote. Ensuring that the voting system remains engaging and relevant is essential.

○ New issues, such as regulating micromobility devices (e-bikes, scooters), managing pedestrian congestion, or ensuring equitable access to public transit, may fill this gap.

2 Potential Inequities in Access:

○ Auto-free cities often rely heavily on public transport. If transit systems are not equitably distributed, certain neighborhoods may feel disenfranchised. A universal voting system could mitigate this by allowing residents to highlight and vote on underserved areas.

3 Increased Complexity in Managing Public Spaces:

○ With more space available for repurposing, cities face the challenge of balancing competing interests. Universal voting systems must be designed to ensure fair representation of all voices, particularly marginalized communities.

4 Adaptation to New Technologies:

○ The rise of micromobility, delivery drones, and other innovations in an auto-free city introduces new governance challenges. Voters would need to navigate these complexities, potentially requiring additional education on emerging technologies.

The Role of AI and Data in Auto-Free City Voting

AI and data analytics can complement universal voting systems by:

- **Providing Real-Time Feedback**: Sensors in public spaces could monitor usage patterns and environmental conditions, offering data-driven insights to voters.

- **Predictive Modeling**: AI tools could simulate the outcomes of proposed policies, helping residents make informed decisions during voting.

- **Equity Monitoring**: Data analysis could identify disparities in service distribution, ensuring that all communities benefit from city policies.

Empowering Communities Through Universal Voting

In an auto-free city, universal voting systems empower residents to shape their environment in meaningful ways. With fewer resources devoted to managing cars, cities can focus on:

- **Sustainability**: Residents can vote on renewable energy projects, waste reduction initiatives, and climate adaptation strategies.

- **Cultural Development**: Universal voting could support investments in art installations, music festivals, and public events that enrich the urban experience.

- **Social Equity**: Voters could prioritize affordable housing, accessible healthcare, and educational programs, fostering a more inclusive city.

Conclusion: The Future of Universal Voting in Auto-Free Cities

The shift to auto-free cities represents an opportunity to reimagine urban governance and civic engagement. By integrating a universal voting system, these cities can empower residents to take an active role in shaping their communities. While challenges remain, the benefits of a participatory, inclusive, and sustainable approach to governance far outweigh the drawbacks. As more cities adopt auto-free models, universal voting systems will play a crucial role in ensuring their success and adaptability.

Chapter 15

Universal Voting Systems and the End of Political Parties

The rise of universal voting systems offers a unique opportunity to rethink how societies make decisions and govern themselves. In particular, these systems have the potential to minimize or even eliminate the role of political parties, shifting the focus toward direct citizen engagement and decision-making. This chapter explores how a universal voting system could transform governance, enable participatory democracy, and ultimately reduce the influence of nation-states.

The Role of Political Parties in Traditional Systems

Political parties have long been central to representative democracy, acting as intermediaries between citizens and their governments. They organize voters, propose policies, and govern on behalf of the electorate. However, this system has inherent flaws:

1 Polarization: Political parties often exacerbate divisions, creating "us versus them" dynamics.

2 Representation Gaps: Parties sometimes fail to represent the diverse interests of their constituents, prioritizing party agendas over public needs.

3 Inefficiency: Bureaucracy and partisanship can slow down the legislative process, delaying solutions to pressing issues.

Universal Voting: A Path to Direct Democracy

A universal voting system could bypass political parties altogether by enabling citizens to vote directly on issues rather than electing representatives to make decisions for them. Key features of such a system include:

1 Direct Decision-Making:

○ Citizens could vote on specific policies, laws, and budgets, eliminating the need for intermediaries.

○ This approach ensures that governance reflects the true will of the people.

2 Online and In-Person Participation:

○ Voting platforms could integrate forums for discussion, where citizens debate issues and share perspectives before casting their votes.

○ Public spaces, such as community centers, could host in-person discussions and town halls to complement online engagement.

3 Real-Time Feedback and Transparency:

○ Blockchain technology and AI-driven systems could ensure secure, transparent voting processes and provide real-time updates on legislative outcomes.

Benefits of Minimizing Political Parties

1 Reduced Polarization:

○ Without parties, voters would focus on issues rather than partisan identities, fostering more nuanced and collaborative discussions.

2 Greater Accountability:

○ Direct voting eliminates the possibility of representatives acting against the wishes of their constituents.

3 Increased Civic Engagement:

○ Citizens would have a greater sense of agency, as their votes directly shape policy decisions.

4 Efficient Governance:

○ Removing party politics could streamline decision-making, allowing societies to respond more quickly to challenges.

Challenges of a Party-Free Universal Voting System

1 Information Overload:

○ Citizens would need access to reliable, unbiased information to make informed decisions on complex issues.

○ AI-driven educational tools and public forums could help address this challenge.

2 Participation Gaps:

○ Ensuring equitable participation across all demographics is crucial. Digital literacy programs and offline voting options would be essential.

3 Risk of Populism:

○ Without intermediaries, there is a risk of policy decisions being swayed by short-term popular opinion rather than long-term considerations.

Universal Voting and the Nation-State

As universal voting systems evolve, they could challenge the traditional role of the nation-state. Key implications include:

1 Decentralization of Power:

○ Decisions could be made at local, regional, or even global levels, reducing the need for centralized national governments.

2 Erosion of Borders:

○ Citizens of different countries could collaborate on transnational issues, such as climate change or trade, through global voting platforms.

3 Redefining Governance:

○ The focus would shift from hierarchical structures to networked systems of decision-making, where power is distributed among citizens.

4 Diminished Role of the Nation-State:

○ As more decisions are made directly by citizens, the need for traditional government institutions could decline, potentially leading to a reevaluation of the nation-state's purpose.

A Vision for the Future

In a fully realized universal voting system, political parties might become obsolete, replaced by a dynamic and participatory model of governance. As technology enables real-time voting and discussion on a global scale, societies could transition toward a more collaborative and inclusive approach to decision-making. While challenges remain, the potential to empower citizens, reduce polarization, and create a more equitable world makes this vision worth pursuing.

By minimizing the influence of political parties and redefining the role of the nation-state, universal voting systems could usher in a new era of democracy —one that is truly of, by, and for the people

About the Author

Dan Paul grew up fishing, playing sports, and building houses in Lake Wobegon. After attending the University of Wisconsin, Madison, and working at the linear fusion reactor lab, he explored alternatives to the dominant codes. While staying in coops, communes, and squats around the country, he learned the fundamentals of setting up and operating progressive, radical, and sustainable alternatives to the codes of the corporations and nation.

Also by Dan Paul

Philosophy of Action Design and Multiplicity by Dan Paul

Book 1 Auto Free Design

Book 2 Workers Health Handbook

Book 3 Save Your Life Prevent Hospital Use

Book 4 Coop Owners Handbook

Book 5 Eye on AI Meeting Needs Sustainably

Book 6 Travels on the Nomadic Terrain

Book 7 Tales of the Urban Shaman

Book 8 Housing in the Danger Zone

Book 9 Corona Time

Book 10 Philosophy of Design, Action and Multiplicity

Landlords Against Eviction

Auto Free USA

AI Frankstein

Forever Flowers Requiem in the Land of Milk and Honey

Websites

https://viaradmedia.org

https://autofreedesign.com

http://workershealthhandbook.com

http://sylphu.com

Social Media

https://www.facebook.com/profile.php?id=61555599616489

https://twitter.com/jonblu

https://www.instagram.com/jonblu61/

Made in the USA
Monee, IL
09 November 2026

67380380R10068